GLOBAL WARMING

Sophie Fern

CONTENTS

A Changing Climate 3

Causes of Climate Change 5

Climate Over Time 13

Scientific Evidence 19

Making Changes 22

Glossary 24

A CHANGING CLIMATE

Weather and climate vary a lot across the earth. The weather in any one given place includes the temperature, wind speed, cloud cover, and amount of rain. The climate is the **average** of many years of weather in that place. For example, the west Texas city of El Paso and its Mexican neighbor, Juárez, are located in the Chihuahuan Desert. Shielded by mountains on three sides, they have a dry climate and more than 200 days of sunshine every year.

We all know that the weather can change, even in a place like a desert, but climate can change, too. If the climate gets warmer—or colder—over a long period of time, scientists describe this as climate change. Right now, the earth's overall climate is getting warmer.

Our planet is about 4.5 billion years old. Over that period of time, it has gone through many changes—really big changes! At different times, our planet's climate has been much hotter and much colder than it is today.

Scientists have given names to different ages in the earth's past. Here is a chart of the earth's climate over the last 2 billion years.

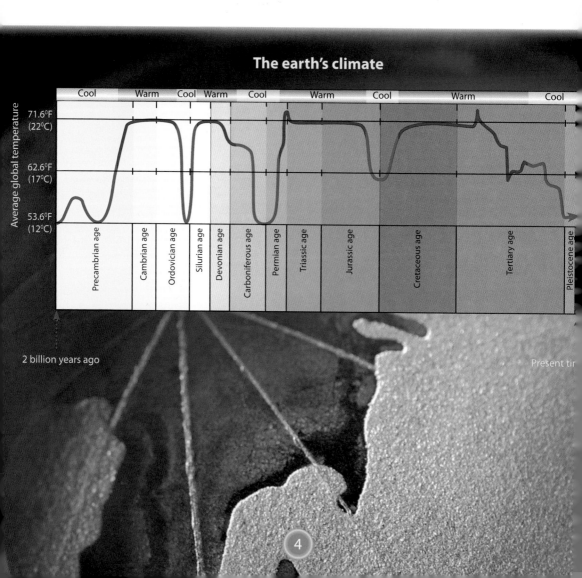

CAUSES OF CLIMATE CHANGE

When scientists measure the age of planets, they talk about billions of years. If you were billions of years old, you'd be a rather interesting person with a lot of history! However, in planetary years, the earth is actually a relatively young planet.

Scientists are still learning about the things that have caused the earth's climate to change in the past. There are a number of factors that scientists think have caused the earth to grow cooler (**glacials**) and warmer (interglacials) at different times.

inter = in between
interglacial = between glacials (between ice ages)

Scientists can gauge the earth's climate changes from the layers in the Greenland Ice Shelf.

> The sun is about 4.6 billion years old. Scientists think that when it is 7 billion years old, it will start to get a lot bigger and its temperature will be cooler—so just as the earth's climate changes, the temperature of the sun can change, too.

The Earth's Orbit

The Serbian astronomer Milutin Milankovitch (1879–1958)

Many scientists agree with Milutin Milankovitch's **theory** that the earth's climate is affected by the way it orbits the sun. Milankovitch noticed that past ice ages on earth seem to be related to the earth's orbit. Milankovitch's theory states that the earth's climate is affected by the following three things:

The Shape of the Earth's Orbit Around the Sun

About every 100,000 years, the earth's orbit changes from being spherical (basketball-shaped) to being elliptical (football-shaped). When its orbit is elliptical, the earth moves farther away from the sun than when its orbit is spherical. When the earth is farther away from the sun, less sunlight reaches it, and it cools down.

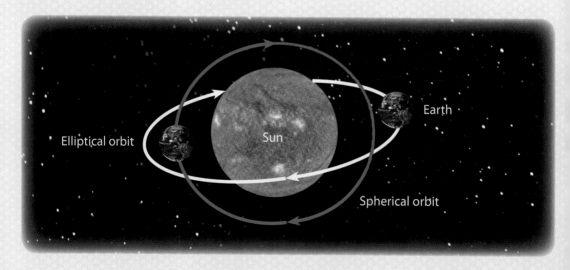

The Earth's Tilt

The earth doesn't sit with north straight up and south straight down—it tilts. The **degree** of tilt varies. Less tilt means that the poles get less sunlight. This makes them icier. The earth's tilt also gives us the seasons.

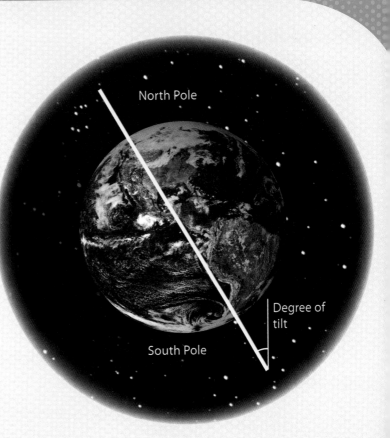

The Earth's Wobble

As the earth goes around the sun, it wobbles. Imagine a spinning top as it slows down. It starts to wobble. The earth's **eccentric** orbit—the way it wobbles—causes the amount of sunlight that reaches the earth to vary.

Scientists are testing the theory that changes in the earth's orbit, tilt, and wobble are linked to the earth's cycle of glacials and interglacials.

Volcanoes

Did you know that volcanoes also affect the overall climate? Volcanoes occur where the oceanic and continental plates—large sections of the earth's crust—meet. Volcanoes also occur where these plates are moving over hot spots—places in the earth's crust where **magma** comes close to the surface.

Lava isn't all that comes out of volcanoes. Ash and gases, such as carbon dioxide, are erupted, too. When volcanoes are particularly active, the greater amounts of ash and volcanic gases entering the atmosphere can **temporarily** affect the earth's temperature. About 65 million years ago, an enormous amount of ash and gas was erupted into the atmosphere by volcanic activity in the Deccan Trapps, in India. Research indicates that when the Deccan Trapps erupted 65 million years ago, 30 billion pounds (13.6 billion kilograms) of sulfur dioxide and 9 billion pounds (4.08 billion kilograms) of hydrogen chloride were released into the atmosphere for every cubic mile of lava that came out!

Scientists wonder whether ash and gas from the Deccan Trapps contributed to the extinction of the dinosaurs by cooling the earth's climate.

Sunspots

Sunspots also seem to affect the earth's climate. Sunspots are dark spots on the surface of the sun that you can't see with your naked eye.

A sunspot viewed close up in ultra-violet light

The sunspots on the sun are colder than the rest of the sun. They are caused by changes in the sun's **magnetic field**, which can reduce the amount of heat energy traveling from the sun to the earth.

The amount of sunspot activity occurs in cycles. When there is more sunspot activity, there are cooler periods on the earth. For example, scientists think that the Little Ice Age, which lasted from about 1250 to about 1850, may have been caused by a period of increased sunspot activity.

The Greenhouse Effect

The earth's **gravity** traps gases next to the earth's surface. These gases, which we call air, make up the atmosphere. If earth had no atmosphere, humans couldn't breathe and couldn't live on earth. The atmosphere also traps the sun's heat energy, making the earth warm enough for humans, plants, and animals to survive.

Some of the gases that make up the earth's atmosphere trap more of the sun's energy than others. Carbon dioxide is one of these. Heat-trapping gases like carbon dioxide are called greenhouse gases.

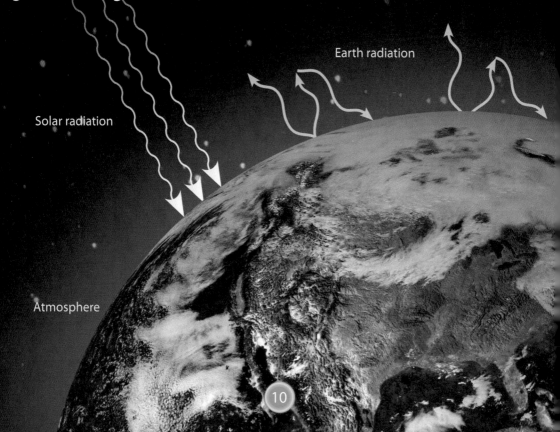

Some scientists have found evidence that increased amounts of carbon dioxide and other greenhouse gases are entering the atmosphere. If this is true, then more of the sun's heat could be getting trapped. This may be causing the earth's climate to get warmer. Research shows that the amount of carbon dioxide in the earth's atmosphere has increased since the 1960s.

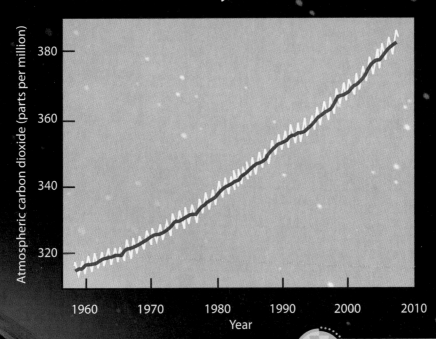

Carbon dioxide in the atmosphere recorded at the Mauna Loa Observatory, Hawaii

Earth

> Without an atmosphere to trap the sun's energy, the earth's temperature range would be like the moon's, which is -387°F to 253°F (-233°C to 123°C)!

All these things—the changes in the earth's orbit, volcanic activity, sunspots, and changes in the amount of greenhouse gases entering the atmosphere—have the **potential** to change the earth's climate, making it either colder or warmer. A change in the earth's climate could affect many things, from where particular plants and animals can thrive to how much of the earth's water is locked up in ice. When less of the earth's water is ice, sea levels rise. When more of the earth's water is ice, sea levels fall.

Did you know that fossil fuels, which include coal, petroleum, and natural gas, were once animals and plants? Animals and plants—and fossil fuels—are made of carbon. That's why they release carbon dioxide when they burn.

CLIMATE OVER TIME

Scientists have found **evidence** that the earth has experienced at least four major ice ages in the last billion years.

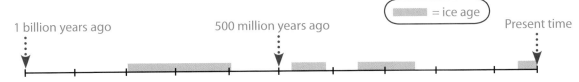

During each of these ice ages, large parts of the world were covered with ice and glaciers. Because parts of the earth—in Greenland and Antarctica—are still covered by sheets of ice, scientists say that the earth is still in an ice age.

At times in the past, much of North America, northern Europe, and northern Asia have been covered with ice. At such times, so much of the earth's water lay frozen on the land that sea levels were lower. Some scientists think that people first reached North America by walking across from Asia. They think that sea levels then were lower than they are today, and Asia and North America were connected by land. Other scientists have different theories.

An artist's impression of how the first people to reach North America may have crossed from Asia

How Warm Is It?

This graph shows variation from the long-term average temperature for the years 1850 to 2007. Since 1980, most years have been warmer than average.

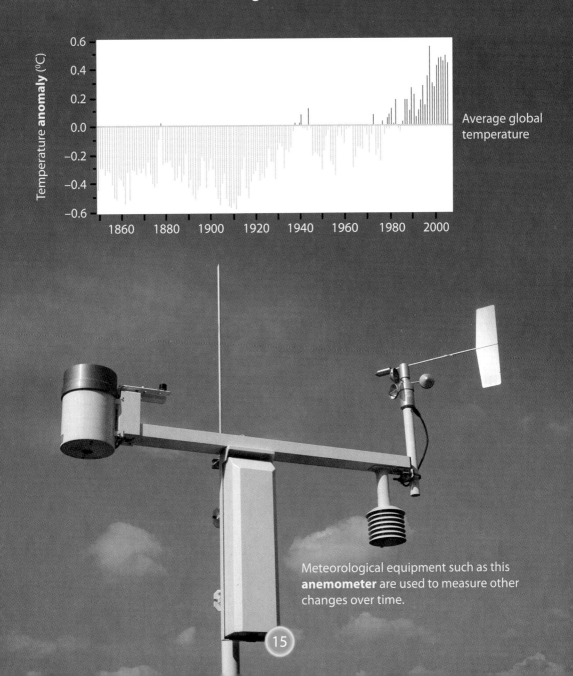

Meteorological equipment such as this **anemometer** are used to measure other changes over time.

Evidence Over Time

Scientists use many different tests to learn about what the earth's climate was like in the past. One test is to look at tree rings. A tree adds a ring to its trunk each year as it grows. By counting the rings on a tree trunk or stump, you can tell that tree's age.

For trees, warm, wet summers are good growing years. In a good growing year, a tree makes a large ring.

Did you know that you don't have to cut a tree down to tell how old it is? Scientists can drill into a tree's trunk and remove a core. This doesn't hurt the tree if it is done correctly.

Corals are another organism that scientists use to learn about the earth's changing climate. Coral reefs are built up by new layers of coral growth, with each layer corresponding to a different year. Scientists remove core samples by drilling into the reef.

Scientists can tell how old coral is by counting the black and white lines. It's like counting the rings of a tree.

Another way scientists learn about the earth's climate in the past is by looking at ice core samples from glaciers and ice sheets. Some glaciers and ice sheets have lasted through several ice ages! Each winter, the snow that falls gets packed down, trapping bubbles of air. Scientists count the layers—just like tree and coral rings—to find out the age of the ice. Then they look at the gas trapped in the bubbles inside each layer of ice. They test this gas to find out what the air was like when the bubbles were trapped.

Air bubbles trapped in an ice core sample

The evidence scientists get from tree rings, coral reefs, and samples of the earth's atmosphere trapped in ancient ice gives us a lot of information about what the earth's climate was like in the past.

SCIENTIFIC EVIDENCE

Scientists agree on how scientific inquiry works. For a hypothesis, or scientific idea, to be accepted, scientists must be able to copy the experiment and get the same results.

- A scientist, or group of scientists, comes up with an idea about our world. This is called a hypothesis.
- The scientist then designs an experiment—a test—to provide evidence that may or may not support the hypothesis.
- The scientist then carries out the experiment, taking good notes about how it is done and what the results are.
- The scientist publishes the results from the experiment and a description of how to do it.
- Other scientists repeat the experiment to see whether they get the same results.
- Even when all this has happened, new results can always disprove the idea.

For example, Dr. Eric Wolff and other scientists have been studying ice in Antarctica, Greenland, and other places. Dr. Wolff has talked about what the ice core samples are telling us.

"It's very exciting to see ice that fell as snow three-quarters of a million years ago."

Looking at the evidence in the ice cores, Dr. Wolff doesn't think we'll have another ice age any time soon, once the one we are in comes to an end.

"However, we may have a heat wave if we are unable to control carbon dioxide **emissions** and other greenhouse gases entering the atmosphere. Our next step is to investigate carbon dioxide in the ice cores, and by understanding what has driven the natural changes seen in the ice record, we will create better **models** to predict how climate might change in the future."

Repeating the research—will this scientist get the same results?

Many scientists are concerned that increased amounts of carbon dioxide and other greenhouse gases in the atmosphere may be causing the earth's climate to warm up faster than expected for the end of a glacial period. They are trying to figure out what effect this might have. For example, could sea levels rise and flood low-lying cities like New Orleans?

If sea levels rise, will New Orleans look like this?

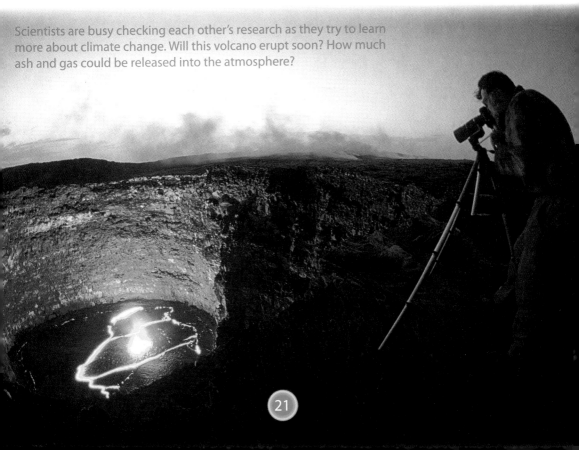

Scientists are busy checking each other's research as they try to learn more about climate change. Will this volcano erupt soon? How much ash and gas could be released into the atmosphere?

MAKING CHANGES

Will the earth's climate continue to warm up? Only by continuing to do research will scientists be able to contribute to the answer to this question.

By studying earlier periods of global warming, scientists are learning about the effects of a warmer climate on the earth. These include a rise in sea level and changes to plant and animal habitats. Scientists are also trying to figure out what happens when such changes happen quickly.

Some scientists believe that big differences can be achieved by lots of people making small changes in their lives. If it makes sense to reduce the amounts of carbon dioxide and other greenhouses gases going into the atmosphere—if this could slow the **rate** of climate change—then what can you do to help?

Here are some changes you could make:

- Walking or biking to school instead of going in a car or a bus
- Using public transportation more often, or carpooling
- Growing plants—because plants take carbon dioxide out of the atmosphere
- Recycling your trash and buying recycled items
- Using compact fluorescent light (CFL) bulbs in your home
- Turning off the lights when you leave a room
- Starting a worm farm using leftover food scraps from your garbage
- Using less hot water when you wash
- Reusing wrapping paper and other items rather than throwing them away

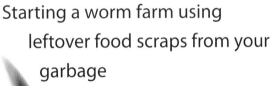

Remember, small changes may make a big difference when lots of people make them!

GLOSSARY

anemometer—a piece of equipment used to measure wind speed

anomaly—a difference from the normal or expected

average—a typical or usual amount

degree—amount

eccentric—departure from the usual path or pattern

emissions—things coming out, such as gases

evidence—data that proves or disproves a theory

glacials—ice ages

gravity—a force exerted by an object, such as a planet, that draws other objects toward it

magma—molten rock inside the earth

magnetic field—an area near a magnetic object in which the magnetic forces caused by the object can be detected

models—computer simulations that can be used to see what might happen

potential—the possibility of being able to do something

rate—speed

temporarily—for a short time

theory—an idea based on scientific evidence that may or may not be true